85 Seconds to Midnight

85 Seconds to Midnight

A Physicist's Argument Against Rearmament

CARLO ROVELLI

ALLEN LANE
an imprint of
PENGUIN BOOKS

ALLEN LANE

UK | USA | Canada | Ireland | Australia
India | New Zealand | South Africa

Allen Lane is part of the Penguin Random House group of companies
whose addresses can be found at global.penguinrandomhouse.com.

Penguin Random House UK
One Embassy Gardens, 8 Viaduct Gardens, London SW11 7BW

penguin.co.uk

First published in Italy by Solferino, 2026
This translation published 2026

001

Set in 12.08/15.21 pt Dante MT Std
Typeset by Six Red Marbles UK, Thetford, Norfolk
Printed and bound in Great Britain by Clays Ltd, Elcograf S.p.A.

The authorized representative in the EEA is Penguin Random House Ireland,
Morrison Chambers, 32 Nassau Street, Dublin D02 YH68

A CIP catalogue record for this book is available from the British Library

ISBN: 978–1–837–31557–4

Penguin Random House is committed to a sustainable future
for our business, our readers and our planet. This book is made from
Forest Stewardship Council® certified paper.

Contents

Preface

Humanity has only one real enemy: our own irrationality. We are afraid of each other, and that's why we massacre each other. Thanks to physicists, we now have a definitive way to slaughter ourselves: nuclear weapons.

This book is a brief history of nuclear weapons, with which we have a good chance of being on the verge of annihilating civilization. It focuses on lesser-known events, which seem to me relevant to understanding the present. What is most surprising in this story is the number of misunderstandings and errors that were the basis of so many cata-strophic political choices of the past. I hope this helps us look more cautiously at the certainties of politicians today.

What animates these pages is the collective bad conscience of my profession, theoretical physics.

We have offered humanity a poisoned gift: nuclear weapons.

My goal is to help avoid what seems to me to be the involuntary but hardly avoidable destination point of the current policy of the governments we have elected: nuclear war.

Carlo Rovelli

1934: Enrico Fermi's Mistake

The experiment that led to the nuclear age was the result of a strange intuition of one of the great scientists of the twentieth century, Enrico Fermi.

In 1934, in Rome, Fermi and his young collaborators were studying the effects produced on various substances by radioactivity, recently discovered and explored by Marie Curie. Irradiating uranium, Fermi had the intuition to try to interpose a piece of . . . paraffin wax. The result, surprisingly, was the transformation of part of the uranium into something else, something that was no longer uranium. The ancient dream of alchemists, to transform one element into another, had come true.

Today we know that paraffin atoms slow down the neutrons that make up radioactive radiation. The slowed neutrons are captured by the nuclei of uranium atoms. These, stuffed with one too many neutrons, lose their stability and break into small nuclei. In this way, uranium is transformed into lighter elements, such as barium and krypton.

We also know that by breaking, like a compressed spring suddenly let free, the nucleus releases a considerable amount of energy, heat. It also releases other neutrons, which in turn can break other nuclei. Which in turn break down, releasing new neutrons and additional energy . . . and so on, engaging in a very rapid 'chain' process which releases more and more energy. In this way, a few ounces of uranium are enough . . . to destroy Hiroshima.

Fermi did not yet know all this, which became clear only in the following years. He had not even understood that the uranium transformation he had observed was due to the break-up of nuclei. He thought, erroneously, that by absorbing neutrons uranium had mutated into new elements never observed before, heavier than uranium itself.

The story that quickly went around the world, scientific and otherwise, was not the correct one: that Fermi had discovered how to break atoms. It was the wrong story of the creation of two new elements.

The Italian Fascist regime swelled with pride at the (erroneous) news of the discovery of the two elements – immediately baptized, to national glory, Ausonius and Hesperium: Ausonia is an ancient Greek name for Italy, Hesperia a poetic name for it.

In 1938 Fermi received the Nobel Prize for this experiment. The citation for the award still mentions this error: 'the discovery of new elements'.

The first to suspect the error and propose the correct interpretation of Fermi's experiment was a German chemist, Ida Noddack, as early as September 1934. She wrote: 'It is conceivable that the bombardment of heavy nuclei with neutrons causes these nuclei to break into several large fragments,' adding 'Fermi himself has shown caution in this regard. However, in a summary of his experiments and in the reports published in the daily press, the [wrong] results are already presented as certain.'

Initially, her clear suggestion was not taken seriously (she is a woman, a chemist, and the mistake is going to be confirmed by a Nobel Prize). But on 19 December 1938, the chemists Otto Hahn and Fritz Strassmann announced that chemical analysis revealed the presence of barium in uranium bombarded with Fermi's 'slowed' neutrons. Barium is an element whose nuclei weigh about half as much as uranium nuclei. It is proof that Ida Noddack was right, and Fermi, without realizing it, had found a way to shatter atomic nuclei.

In January of the following year, the physicists Lise Meitner and her nephew Otto Robert Frisch clarify the theory of the process. Frisch calls the process 'fission', by analogy with biological fission: the separation of a living cell into two cells. In a second work shortly afterwards, the two physicists show that, when an atomic nucleus is hit by a neutron and breaks, it in turn emits neutrons. It becomes clear to many that these neutrons can set off a chain reaction. From a small amount of uranium or plutonium it is possible to build a bomb of immense power.

Fermi's experiment is therefore even more important than the Nobel Prize citation acknowledged: it is the discovery of a new source of energy. A great gift for humanity.

But the gift is too great. A small bit of uranium can release enough energy to demolish cities, burn

alive millions of human beings and destroy civilization itself. A poisoned gift.

What to do with the ring of power? Knowledge is a
deadly friend – sang Greg Lake – if no one sets the
rules.

1941: Two Geniuses in Semi-secret Conversation

Somewhere between 15 and 21 September 1941 (we do not know the exact date), two men, linked by a close scientific and personal relationship, met in Copenhagen. Books, plays and documentaries were subsequently written, and friendships broken, because of this meeting, because of the words exchanged that evening. Perhaps that dialogue changed the fate of humanity.

What exactly the two protagonists said to each other no one knows. I think they didn't remember it either, years later, trying to reconstruct the words of that day. Memory is unreliable; it often rearranges things as it suits us. The two never agreed on a common version.

One of the two was Niels Bohr. He was fifty-five

years old and recognized as the father of quantum physics. He had received the Nobel Prize in 1922. He was a cultured, complicated, profound, affable Danish man. He directed a research centre in Copenhagen that drew the brightest minds of the new physics.

The other was Werner Heisenberg. He was younger – thirty-nine years old – and had received the Nobel Prize in 1932. He was a simple, direct young German, perhaps a little naïve. He was the one who had found the key to calculating quantum phenomena. They were two of the greatest scientists of the twentieth century. Perhaps two of the greatest scientists ever.

The affection between them was deep and sincere. Heisenberg had visited Bohr's institute in Copenhagen when he was still very young, and Bohr had immediately realized the exceptional qualities of the German student. They had walked together in the mountains, talking about science and philosophy. Theirs was an intense human and scientific relationship, which had led Heisenberg, then just twenty-three years old, to the main conceptual leap towards the equations of the new mechanics, quantum mechanics, destined to completely revolutionize modern science, perhaps even modern thinking itself.

Heisenberg revered Bohr as his teacher. Bohr saw in Heisenberg the young genius who had completed the great revolution of thought he had started and nourished: quantum mechanics, the basic theory that allows calculations to be made to predict and understand the atomic and nuclear phenomena at the heart of atomic energy.

But this affection, esteem, veneration, between the master and his outstanding disciple were put to the test by the circumstances of the 1941 meeting.

The international political climate had heated up in the previous decades (as it has today). The First World War had destroyed the great optimism about 'progress' in the *belle époque* and had allowed the communist victory of the Russian Revolution, which in turn had kindled hopes of socialism, justice and class equality throughout Europe. The furious reaction against these hopes had led to the victory of fascism in Italy, then of Nazism in Germany, and had unleashed the ferocious Spanish Civil War, where the democratically elected Republic was supported by Soviet Russia, while Generalissimo Franco, who would later keep Spain under a brutal fascist dictatorship for decades, was directly supported by Nazi Germany and Fascist Italy. France and England, for fear of socialism, did not go far in defending the Spanish democracy: between fascism and communism, the ruling elites always end up seeing fascism as more convenient.

With the permission of the great powers, Germany, which had been defeated in the First World War, armed itself (as it is arming itself today). Under Nazi leadership, which presented its historical mission as anti-communist, it was seen by many as the only force capable of stemming and overthrowing Soviet communism. Russia feared (as it fears today) the rearmament of Germany and (as today) sought security in the control of its neighbours. Germany, crushed after the First World War, and now rearmed, had demanded and obtained border adjustments along cultural and linguistic lines (as happened not long ago after the collapse of the Soviet Union), and annexed German-speaking peoples, generally consenting, in Austria and part of

Czechoslovakia. For the same reason, it demanded part of Poland.

Stalin pressed for an anti-German alliance with France and England, but they showed little enthusiasm: it would have taken much more for them to accept the step of allying themselves with communists against fascists. Poland (as today) did not even want to hear about alliances with Russia.

After the failure of a last attempt at British mediation between Germany and Poland, Stalin and Hitler surprised everyone by signing a non-aggression pact and an agreement for the partition of Poland. The Nazi–Soviet alliance seemed geopolitically unthinkable. But, as always, it is not ideology that prevails but realpolitik. Stalin avoided finding himself with all of Europe against him, and Hitler found the door open to take France by surprise and avenge a defeat that still burned.

Europe is on fire.

By 1941, Germany has triumphed everywhere. Hitler looks like the new Alexander the Great. In London there is discussion about whether to end the war and ally with him. Winston Churchill must fight hard to continue the conflict and pursue the strategy with which England has ruled the world for a century: to push others to go to war against each other. Russia, frightened by Hitler's unexpected continental triumphs, rushes to arm itself (as it is doing now). Germany, worried (as now) by the abrupt acceleration of Soviet rearmament, and blinded by its own triumph, decides that the window of opportunity could close soon, and does what many wanted it to do and what was planned from the beginning: it invades Russia.

It is in this red-hot climate, at this time when Europe is in flames, that physicists glimpsed the possibility of creating hell on earth. And it is in this climate that the two leaders of the new physics meet, in German-occupied Denmark: Niels Bohr, the Danish master, and Werner Heisenberg, the German student.

What did they talk about? Who will win the war? Nuclear reaction? The super-bomb? Who will have the bomb? Whether or not to build such a bomb in the middle of a ferocious war? How close is the possibility of building it?

Afterwards, Bohr will take to the United States a sketch made by Heisenberg on that occasion.

He will say that it represents a simplified diagram of the bomb. That is not true. At most it represents a sketch of a peaceful reactor.

Heisenberg will say that he wanted to talk about the moral question: to make or not to make the super-bomb. Bohr will say that that was not true.

I don't think even they, in the post-war period, knew what they had really said to each other, in the dramatic context of an occupied country, and in the difficult reversal of roles into which history had forced them: the teacher, now part of a crushed nation; the pupil, now among the invaders triumphing everywhere.

What exactly they told each other we will never know. But what we can perhaps say is that, together, the two great scientists were not up to the task. They did not say the things that could have been said. They didn't listen to each other, they weren't lucid.

What they might have said to each other, if better angels had been sitting on their shoulders, is something that we can imagine – and perhaps the history of the world might have been different. We will get to that at the end of these pages.

1942: Why Didn't Germany Build the Bomb First?

Europe plunged into the abyss of the world war the year after Fermi's Nobel Prize. The idea that the nuclear fission he had discovered could be used to build a super-bomb soon spread across the rarefied world of atomic physics. The obvious question was whether the super-bomb could be built and used in time to affect the current war.

What happens at this point determined the history of nuclear weapons and the dramatic situation of risk in which we find ourselves today: and it was a tragic misunderstanding. This book, as we shall see, tells a story in which a series of misunderstandings and misinterpretations led to a sequence of catastrophes.

The first misunderstanding to happen is as

follows. The Americans threw themselves headlong into a colossal effort to build the bomb in a short time, with the explicit motivation to build it faster than Nazi Germany and the firm conviction that the Germans were also rushing to build it. But it was not true: the Germans were not at all seriously engaged in building the bomb. Why?

German science was the most advanced. Germany had the best scientists. The German industrial system was very powerful. It had uranium mines in Czecho-slovakia, and the so-called heavy water (which does more or less the work of Fermi's paraffin, namely it slows down neutrons). While in the United States (and Russia) it was scientists who repeatedly alerted politicians to the possibility of the super-bomb, in Germany the political leadership, more attentive, had realized very early that there could be the possibility of a deadly atomic weapon: it did so more than a year before the American leadership. Despite this considerable starting advantage, how did the Germans not even come close to having the bomb?

The reason is simple: they did not make a serious commitment to building it because the choice *not* to commit to the construction of an atomic bomb was the rational choice at that time. This is because there was no realistic possibility of building the super-bomb in a time short enough for it to influence the course of the war. The priority was to win the war. The time to think about the super-bomb was later, at the end of the war.

That this was the rational choice has been confirmed by history: the Americans invested colossal industrial, intellectual, financial and military resources into the construction of the super-bomb, but by the time they had finally managed to make it work, Germany's complete capitulation had already taken place. Peace treaties had been signed. The crazy showdown that had led to 70 million deaths was already decided.

The reasoning behind Germany's decision not to engage in the construction of the super-bomb still generates some heated debates that raise strong emotions among scientists. We will see why. People talk about hypothetical scientific errors, moral issues, hidden motivations. These are empty discussions, because the facts are transparent: the German choice was the rational choice, and the conversation and decisions about it are historically well documented.

Nazi Germany had seriously considered the possibility of developing nuclear technology. Several research projects had been funded. The *Uranverein* (Uranium Society) project began as early as April 1939, a few months after it was clear that Fermi's experiment had produced nuclear fission. The *Uranverein* ended almost as soon as it had begun, with the enlistment of many German physicists into the Wehrmacht. But in September 1939 the *Uranprojekt* (Uranium Project) was initiated, and it took three directions: the development of *Uranmaschine*, a nuclear reactor (like the one Fermi proposed to build in the US); the production of heavy water; and the separation of uranium isotopes (only one of the isotopes is good for the bomb).

Werner Heisenberg, who Fermi had studied with and befriended, and his colleague and dear friend Carl Friedrich von Weizsäcker (perhaps the one who had pushed Heisenberg to go and talk to Bohr in 1941), were at the centre of these projects. Repeatedly questioned by the War Ministry of the

German government, they had replied that it was not reasonably plausible to build the bomb within the short time the war was expected to last. Historical documents clearly confirm this fact. Hypothetical political, ethical or other motivations, as I have mentioned, have been much discussed, but inappropriately. One fact cannot be doubted: the physicists were, quite simply, right.

Heisenberg recalled that in 1939 physicists concluded that 'in principle atomic bombs could be built, but it would take several years,' certainly no less than five. He later claimed that he 'did not report it to the Führer until two years later, and very carelessly', because 'I did not want the Führer to be so interested as to immediately order great efforts to build the atomic bomb. Speer thought it was better to abandon the idea, and the Führer also reacted in this way.' Albert Speer was the Minister for Armaments and War Production.

When Reich Field Marshal Erhard Milch asked explicitly how long it would take the United States to develop a nuclear weapon, he was given 1944 as the very minimum estimate. That was correct: the US got the bomb in July 1945.

Details aside, it is clear that the German military correctly judged that nuclear fission would not contribute significantly to the war it was fighting. In the delicate management of resources in times of war, it could therefore not be a priority.

In January 1942, the army handed over the programme to the Kaiser-Wilhelm-Gesellschaft, the civilian research institution which after the Second World War became the Max-Planck Institute, still today the public institution for science in Germany. The nuclear project retains its *kriegswichtig* designation ('of war importance'), and funding continues to come from the military, but the funding is tiny and restricted to areas of uranium and heavy water production, separation of uranium isotopes and *Uranmaschine*.

It is enough to look at the intellectual output of the German theoretical physicists during the war, for example that of Heisenberg, to understand that German scientists had no passion for, no dedication to, the construction of the bomb. They concentrated on other questions.

In the final months of the war in Europe, a special Allied operation called 'Operation Alsos' was charged with reaching German scientists and capturing material relevant to nuclear research. Among the German scientists who had published reports as members of the *Uranverein*, the Americans captured Erich Bagge, Kurt Diebner, Walther Gerlach, Otto Hahn, Paul Harteck, Horst Korsching, Carl Friedrich von Weizsäcker and Karl Wirtz. Werner Heisenberg voluntarily surrendered in May 1945. The Allies were shocked by how little they found.

The bomb was made in wartime in the US and not in Germany for the simple reason that the Germans did not try, while the Americans, as we will see in the next chapter, threw themselves headlong into it.

1942: The Manhattan Project

On the other side of the game of massacre, in the United States, the government, unlike the more shrewd Nazi government, had not had an inkling of the possible super-bomb. It was physicists – my people – who repeatedly insisted on two points with the American government. First, that it would be possible to build the bomb before the war with Germany was resolved. And second, that it was necessary to do it, and quickly, because the super-bomb was almost certainly being built by the Germans. They were wrong on both counts.

Repeatedly urged by scientists, the US government finally reacted. It decided to engage heavily in the effort to build the super-bomb. This effort was called the Manhattan Project, and it led to the construction of atomic weapons during wartime. The name 'Manhattan' comes from the American military habit of naming operations after the city that hosts the headquarters (Manhattan, New York, in this case).

In size, the Manhattan Project was colossal. To understand just how large it was, it was, it is enlightening to compare the super-bomb efforts of Germany and the US. In terms of financial and human resources, the contrast between the *Uranverein* and its derivatives in Germany and the Manhattan Project in the United States is striking.

The German *Uranverein* had a budget of 8 million Reichsmarks, equivalent to about $30 million in 2026. At most, a few dozen scientists worked on it. The American Manhattan Project directly employed more than 130,000 people and indirectly involved about 500,000 – almost 1 per cent of the entire US civilian workforce. It cost about $2 billion in 1945, or $30 *billion* in 2026. Which means that the US economic effort for the super-bomb was about a *thousand times* that of Germany.

This effort did *not* contribute in any way to winning the war against Germany. Germany surrendered months before the bomb was ready. If we really want to say that it influenced the war it would be because it diverted resources from the actual war effort, slowing down victory.

The rational choice for winning the war was therefore obviously to focus on winning it with the available weapons, and then possibly to invest resources in the bomb later. In other words, regardless of any controversy about the reasons (to which I will return), the German choice not to chase the super-bomb to win the war was simply, I repeat, the rational choice.

With the two erroneous arguments I mentioned – namely, that the bomb could be built before the end of the war (false) and that Hitler would produce it otherwise (false) – physicists convinced US politicians to shower them with money. Scientists such as Einstein, John Wheeler, Edward Teller, Leo Szilard, Bohr himself, and others, were grossly wrong in judging what German science was doing. The fear of a Hitler super-bomb was unfounded. But it was this fear that motivated the huge effort of the United States.

If scientists in the United States had not panicked, if they had waited, if they had not made a mistake in deducing what their colleagues in Germany were doing, if they had simply talked to each other (they were close friends, before the war), nuclear research could have developed in a moment of peace.

Why is this crucial with respect to today's situation? Because if the atomic weapon had been born in peacetime, everything could have been different. The force of the atom could have been a force for peace, and human beings could perhaps have been reasonable enough to submit it to shared international control, as they have been able to do with biological weapons or chemical weapons. Perhaps we might not live in the nightmare we are in today, the constant and imminent risk of being burned alive. Sometimes, human beings know how to be reasonable and act in concert for the good of all.

But the bomb was built during a terrible war, motivated by the unfounded fear that 'the enemy' would produce it. The Americans threw themselves headlong into building the ultimate weapon of mass destruction with the wrong motivation. It is baseless fear that triggers our worst aggression.

We are in the same situation now. The reason why we seriously risk, today more than ever, being burned alive, is that we have an exaggerated fear of each other. It is fear, unmotivated and irrational fear, that (then as now) makes us aggressive.

We hurt ourselves for fear that our enemies will hurt us. We are afraid of our own shadow. We frighten each other, and this triggers wars. Could we be any dumber?

It was the same in the years leading up to the First World War: France armed itself out of fear of Germany, Germany armed itself out of fear of France. England built more battleships because Germany did, and then Germany built them bigger, and so on, until the inevitable outbreak of war. The slogan, then as now, was 'If you want peace, prepare for war.' Preparing for war has always led to war, not peace.

It is the same today. Israel is afraid of Tehran and bombs it. Tehran is afraid of being bombed by Israel and the United States and wants nuclear weapons. Japan is afraid of China and is rearming. China, surrounded by American military bases, is afraid of being suffocated by the United States and its allies, and arms itself. The United States is afraid of China's growth and increases military spending.

Russia is afraid of NATO and invades Ukraine. Europe is afraid of Russia and is arming itself.

And so on, ad infinitum, in a whirlwind of mutual fears that constantly lead us to massacre one another.

What if we stopped? Wouldn't it be better for everyone?

It is not utopia. Within countries, we have managed, more or less, to find a way not to constantly kill one another. Is it possible that we are so stupid that we cannot do the same *between* countries?

1945: Hiroshima and Nagasaki – Why Use the Bomb after the War Has Been Won?

Nuclear weapons have been used in war only twice, within a few days of each other. Both times by the United States. The decision that led to using them was motivated by a serious error of judgment, plus simple thirst for power.

The war was ending. In Europe, Italy had long since surrendered and the German divisions had been destroyed (in large part, at enormous human cost, by the Red Army of the Soviet Union). Germany, finally, had also surrendered. In Asia, Japan had already been driven out of many of the territories it had occupied during the war, partly by the Chinese and partly by the Americans.

Japan was trying to negotiate a dignified surrender.

Most wars in history have ended when one of the contenders accepts the inevitability of surrender. A price is paid; peace is negotiated. Japan explicitly asked Russia to help it negotiate a dignified surrender with the United States.

The situation between Russia and Japan during the war is rarely discussed in the West. Russia and Japan were not at war with each other. The 1905 conflict, won by Japan, and then undeclared conflicts in Mongolia and Manchuria in 1938 and 1939, which ended in crushing Japanese defeats, had convinced Japan that they should sign a non-aggression pact with the Soviet Union. This pact allowed Russia to concentrate on defending itself against German invasion in Europe and allowed Japan to concentrate on conquering vast territories in Asia.

The Americans and the British were pushing for Russia to exit the pact and attack Japan. Japan feared exactly this, that Stalin would betray it and attack it.

In 1945, in the midst of the difficulties with a war already obviously lost, Japan tried to convince Stalin to help it reach a negotiated end to the conflict with the United States. The argument Japan uses is transparent: the future lies ahead with a world divided between the Soviet Union and the United States; it is better for Russia to have a friend in Japan rather than a Japan that is a vassal of the US (as did indeed happen later).

But Stalin thought he had a better option. Roosevelt and Stalin came to an agreement. At the Tehran and Yalta conferences, Roosevelt obtained from Stalin the promise, kept secret, to attack Japan after the capitulation of Germany. Indeed, precisely three months after the capitulation of Germany.

Stalin behaves like a gentleman. He announces to Japan in advance that he is withdrawing from their non-aggression pact, and fulfils his promise to Roosevelt to the letter: he attacks Japan exactly three months (to the minute) after Germany's capitulation. After the war in Europe, Stalin moves troops to the East and masses them on Russia's eastern border. On the night when the three months expires, he invades Manchuria, a vassal state of Japan where the best part of the Japanese army is stationed.

The agreement between Roosevelt and Stalin is that Russia and the US should invade at the same time. Japan would end up like Korea and Germany:

divided in two by a half-American, half-Soviet occupation.

This would have meant that both Berlin and Tokyo would eventually be in Soviet hands. Only Rome among the capitals of the Axis powers would fall into American hands. The Soviet Union, which had annihilated most of Hitler's military strength, would emerge with an enviable geopolitical position. This must not have pleased London and Washington, since Soviet communism, which they had tried so hard to prevent since the Russian Civil War, was basically the central problem from the beginning. In Churchill's repeated words, Bolshevism was to be 'strangled in the cradle'.

But at the Potsdam Conference, at the end of July 1945, the new American president, Harry Truman, receives the news of the 'Trinity test', the first test of an atomic bomb in the New Mexico desert. The Manhattan Project has finally been successful; the United States now has the super-bomb. And Truman changes strategy.

He says nothing to Stalin but decides to prevent the Russian invasion of Japan. The super-bomb allows him to avoid two things: not only a negotiated surrender of Japan, but also Tokyo falling into Soviet hands. This, I believe, is the first reason to use the bomb: to get ahead of the open question of who will dominate the world after the end of the war. To avoid not only a dignified surrender of Japan but, even more so, communists in Tokyo.

So: first, no dignified surrender for Japan. A mere surrender by Japan is not enough for the American government; it wants unconditional surrender. It wants total victory. Second, no joint invasion of Japan. It does not want to share the triumph with the Russians. The Americans did not want Russia to be the power that reached Tokyo.

Today, it may be hard to remember how different the world looked in 1945. The number of German divisions fighting against the Americans and British in Western Europe was negligible compared to the number trying to stop the Soviet counter-offensive after the battle of Stalingrad. Russia was the real power that had destroyed the German army.

The war had been won to a large extent by the Soviet Union in Europe, and partly by China (which would soon become communist as well) in Asia. Out of every ten German soldiers killed, eight had been killed by the Russians. It is a macabre calculation, but in a world where decades of propaganda has almost convinced the Italians and the French that Nazism was overthrown by the Italian and French resistance, and in which Russia was not even invited to the Western celebrations for victory in the Second World War, perhaps it is not useless to remember some figures, even if they are ones of horror.

But there is an additional reason for the atomic bombs on Japan, I think even more relevant to the story I am telling. Truman was convinced that the overt use of the bomb would put his country in a position to become the sole nuclear power, and thus the world's only superpower.

The spectacular explosion of the atomic bombs on Hiroshima and Nagasaki was first and foremost an immense demonstration of power. The huge effort that the Manhattan Project had required had led Truman to think that only the United States could accomplish it. He was convinced that the Soviets could never do the same. The two bombs on Japan would credit the United States as the sole nuclear power *for ever*. This, I believe, was the real reason for burning alive 200,000 men, women and children in Hiroshima and Nagasaki. The scream of the gorilla beating its chest and telling the forest that it is the strongest.

But calculations, once again, were wrong. The Russians built the super-bomb shortly after, in a very short time. Truman burned alive hundreds of thousands of men, women and children for another miscalculation. This is how the atomic age began.

Do not take my previous comments as anti-Americanism. Far from it. I'm not complaining about the stupid aggression of the United States: I'm complaining about stupid human aggression. In history, even in recent history, we have witnessed expressions of astounding violence by all sorts of governments. In the past we have witnessed worse.

I love America, which has played an important role in my life. I lived in the US for many years, and I respect and admire it for so many aspects of its culture and civilization. But it is not a model of moderation; despite what its propaganda and most of its citizens insist, it is no better than the rest of the world. Allow me to remind you of a few details, because they are relevant for this story:

It is a country founded upon a series of genocides, which have virtually wiped out all the populations and cultures of the part of North America it now occupies. It got rich because of the widespread use of slavery, because of the slave trade from Africa. It massacred itself in one of the bloodiest wars in history, the American Civil War; even today, it has one of the highest homicide rates among high-income countries. And it has not hesitated to slaughter hundreds of thousands of human beings, for strategic geopolitical calculations, in an almost uninterrupted series of more than thirty conflicts throughout the so-called post-war period, which have led to a number of deaths that can be estimated at between 5 and 10 million. All were wars in which the United States was the aggressor, never the attacked.

I'm not saying it's a country worse than others. It is not. What I want to say is that, for a country like this, it must not have been so difficult, towards the end of a global massacre of 70 million, to decide to burn alive a few hundred thousand more men, women and children in Hiroshima and Nagasaki. The bombing of Tokyo, with conventional weapons and napalm, shortly before had already caused more deaths than an atomic bomb.

I am not even broaching the general moral question, except perhaps by quoting a sentence pronounced after the war by the American general Curtis Emerson LeMay, responsible for those bombers: 'I imagine that if we had lost the war I would have been tried as a war criminal.'

But don't get me wrong. Don't take the crudeness with which I described these political decisions as cynicism, or hopeless pessimism.

It is the opposite. I think that humanity, although it grew up on wars, exterminations, genocides, slavery, rapes and nuclear massacres, is the same humanity that has been able to reduce violence, abolish slavery, agree on systems of laws and enforce them. The same humanity that learned to live together with millions of others, fight poverty, re-distribute wealth, fight privileges, spread well-being and education, and so much more.

It is the opposite of cynicism and pessimism that motivates these pages: it is the conviction that humanity, intelligence and reasonableness have often prevailed and still have a chance to do so.

But it is only one possibility. A possibility that requires, I believe, commitment, and not indifference, from everyone. All of us and each one of us. Politicians, journalists, newspaper editors, intellectuals of all disciplines, citizens who vote, people of all kinds who know how to see the common good beyond partisan interests. Seeing others not as enemies to be feared, but as members of the same global collective who must find a way to coexist.

Those who know how to write, please write; those who know how to organize, please organize; those who know how to convince, please convince; and those who have to make decisions, please think about the long term, for the good of all.

1949: The Russian Atomic Bomb

As soon as he learned of Hiroshima and Nagasaki, Stalin accelerated the Russian nuclear programme, which already existed.

The Russian physicist Georgy Flyorov had suspected that the Allied powers were secretly developing the bomb as early as 1939. In April 1942, he addressed two letters to Stalin: 'The results [of the bomb] will be so overwhelming that it will not be necessary to determine who is to blame for the fact that this work has been neglected in our country,' adding: 'It is essential to manufacture a uranium bomb without delay.' Stalin immediately recalled Russian scientists from military service and began a project to develop the bomb under the guidance of the physicists Anatoly Alexandrov and Igor Kurchatov.

Early efforts consisted mainly of working at Laboratory No. 2 in Moscow and gathering information on the Manhattan Project from Soviet spy networks. In the final months of the war in Europe, a Soviet task force competed with the Western Allies' 'Operation Alsos' to reach German scientists and capture material relevant to nuclear research. The Soviet nuclear project immediately used the uranium mines of Taboshar in Tajikistan, the deposits in Czechoslovakia and the heavy industry of East Germany to extract and refine the uranium, and for the production of tools (further confirmation of Germany's resources).

Lavrentiy Beria himself, then People's Commissar for Internal Affairs of the USSR, was put in charge of the atomic project in December 1944. Priority was given to replicating the plutonium implosion weapon that the Americans used in Nagasaki. The first Soviet nuclear test, the RDS-1, (in Russian, РДС), took place in Semipalatinsk in Kazakhstan on 29 August 1949. Codename 'First Lightning' (Первая Молния, or Pervaya Molniya).

Only four years after Hiroshima and Nagasaki,

Truman's plan to make the United States the sole nuclear power had already failed.

Several Western scientists moved to Russia to contribute to Soviet communism. Among them was the Italian Bruno Pontecorvo, one of Fermi's first collaborators, who reached Russia in 1950 and worked until his death in what is now the Joint Institute for Nuclear Research (JINR) in Dubna.

An interesting example of collaboration with Russia is that of the German physicist Klaus Fuchs. Fuchs worked on the American bomb, made important theoretical contributions to its construction, was a friend of Richard Feynman, and passed information to the Russians. He was motivated not by personal gain but by his conscience. The experience of the Nazi regime made him profoundly opposed to war. He could not understand the West's reluctance to share atomic secrets with the allied Soviet Union. He sensed the danger of having one country rapidly rise to absolute power thanks to the superbomb and, most importantly, was not willing to let the United States annihilate Russia.

*

Years later, an American journalist asked one of the scientists who worked on the bomb in Russia if he felt guilty for helping to give Stalin the atomic bomb. The answer is enlightening, and I think it should be taken seriously today. The reply was that for years the scientist could indeed not sleep at night, but this was *before*, not *after*, the day they told Stalin that the bomb was made: at night he had nightmares in which he dreamed of millions of men and women in Leningrad and Moscow burned alive by American atomic bombs. When the Soviet Union finally got the atomic weapon, he thought that at least for the moment all those people were safe, and he could finally sleep peacefully.

Today, Russia possesses just over 4,000 nuclear warheads, the largest number of super-bombs in the world. Russia also has about 1,700 missiles to launch the warheads, the largest confirmed strategic arsenal on the planet.

Yet, in Europe, particularly in some countries, instead of finding solutions to current disagreements, there is a burning and foolish desire to wage war on Russia.

1950: The Balance of Terror – The World Saved by Indiscipline

With the Russian bomb, the world enters the era of so-called nuclear deterrence.

This means that two countries, armed to the teeth, can totally destroy each other in a matter of 15 minutes, just by pressing a button.

The balance works like this: each has enough sensors to detect the other's missile launches and has some time, after detection, to unleash their own complete destructive power. But this time is thoughtlessly short. From the time of detection, each of the two parties has a few minutes available to destroy their opponent in turn. It is called 'MAD' – 'Mutual Assured Destruction'. The abbreviation says it all.

It is a delirious poker game at the last breath, as what is at stake is the survival of all of us.

To understand the risks, it is useful to understand how exactly this works.

It is a matter of a calculation of minutes. If one of the two countries, the United States or Russia, launches a nuclear attack, ballistic missiles take a certain number of minutes to fly to their targets. These minutes are everything. During these minutes, the attacked country has time to detect the attack and communicate the news to their political leaders, who then have (a very short) time to give the order to launch the retaliation. Retaliation can completely annihilate the country that attacked first. The optimistic idea is that in this chilling situation, no one drops the first bomb.

All this would happen in a few essential minutes. For example, the President of the United States has between 3 and 7 minutes to decide to retaliate, from the moment he receives the news, before the atomic bomb annihilates him. This is the insane balance of terror in which the world has lived for decades and is still living, in today's growing belligerence.

★

The balance is precarious and often staggers. Leaders of the Soviet Union and the United States repeatedly negotiated to prevent it from breaking, and to re-establish deterrence, often walking a tightrope.

An example of this infernal whirlwind is the Cuban crisis in October 1962, an event that allows us to better understand the current war in Ukraine. The details are as follows.

First, the Americans place nuclear missiles in Turkey. Turkey (like Ukraine today) is too close to Russia, and this may make Russia's reaction time insufficient, thwarting the logic of deterrence. Russia reacts by sending ships to install atomic missiles in Cuba, in order to balance things out. The United States, symmetrically, cannot afford to have hostile nuclear missiles so close, because it would leave no time for reaction. So the US threatens to sink the Russian ships delivering the weapons to Cuba. The Soviet Union threatens to react with a nuclear attack. Neither side wants to give in, and the world marches towards catastrophe. Advisers in both Washington and Moscow are pushing their respective presidents to go ahead with this cursed poker. Only the combined nerve of the two Kennedy brothers on the one hand, and Nikita Khrushchev on the other, managed to avoid a catastrophe that seemed unstoppable. On the last possible day, the Kennedys and Khrushchev agree to both take a step back: the Russian ships stop and turn back, while the Americans in exchange agree to remove the missiles from Turkey. The Soviet Union thus achieves its goal of removing the American missiles from Turkey. The detail that makes the agreement possible is that the withdrawal of missiles from Turkey will not be announced immediately, so that the young President Kennedy can save face internally. Thanks to this ruse, the balance of terror keeping the atomic catastrophe from happening is restored.

In this precarious balance, the risk of a false alarm is very high. There have been numerous false alarms, and total nuclear destruction has been a hair's breadth away several times. Sometimes because of simple technical or communication errors.

During the Cuban crisis, a Russian nuclear submarine had concluded from the incomplete information it had at its disposal that nuclear war had begun. The captain had interpreted explosions heard around it as bombs that an American ship was dropping on them. The captain concludes that he has to give the order to launch the nuclear warhead. Soviet protocol for that submarine requires (uncommonly) that the three officers at the top of the submarine's command have to approve the launch (usually there were only two). The two highest officers approve.

The third officer, Vasily Aleksandrovich Arkhipov, opposes it. On the submarine there is an altercation; Arkhipov does not behave as he should. The world has been saved.

Arkhipov later received top honours both in Russia and in the West. Thomas S. Blanton, director of the US National Security Archives, called Arkhipov 'the man who saved the world'. The world had been saved by his lack of discipline. I sincerely hope that everyone can have this lack of discipline.

On 26 September 1983, Colonel Stanislav Petrov refuses to launch a nuclear response after a radar sighting of incoming American missiles. Patient, he checks minute by minute, waiting, contravening orders: he does not notify his superiors. His wisdom saves us from nuclear holocaust. The American missiles turn out to be an error of the Soviet radar system. Once again, indiscipline saves us from catastrophe.

On 9 November 1979, computer errors at Peterson Air Force Base, in the United States, give the alarm and trigger preparation for the reaction to a (non-existent) large-scale Soviet attack. The army warns National Security Advisor Zbigniew Brzezinski that the Soviet Union has launched 250 ballistic missiles with a trajectory towards the United States. The president must make the retaliatory decision within 7 minutes. Computers indicate that the number of incoming missiles is 2,200. Nuclear bombers prepare for take-off. Six to seven minutes before the launch of the response (i.e. the obliteration of Russia from the planet), the PAVE PAWS satellite and radar systems warn that the attack is a false alarm. The world, this time, is literally saved by a minute. Next time?

Today, the balance has become more unstable, for three reasons.

The first is that the risk of a false alarm continues to be present. It's just a risk, but time passes. It's like keeping on rolling two dice always hoping that two sixes don't come up, trusting that the probability of them coming up is low. Of course, it *is* low, but sooner or later two sixes do come up. Sooner or later, a false alarm is not stopped by an unruly Vasily Aleksandrovich Arkhipov.

The second reason for instability is that, today, there are three major nuclear powers, no longer two, and belligerence is growing. The powers are frightening each other, and the balance is increasingly precarious. We have heard repeated nuclear threats in recent years and months, obviously from those who feel weaker. Those who feel stronger do not need to threaten.

The third reason, and I believe the most serious one, is the fact that things are constantly changing, and technological progress is constant and unstoppable. As soon as, for a moment, one of the contenders thinks it is equipped with enough technology to be able to be faster than the counter-offensive, the temptation to destroy enemy nuclear missiles (and a few million human beings as collateral damage) is irresistible. Every technological advance – today, for example, artificial intelligence – risks altering the precarious balance and throwing us into the nightmare.

The *Bulletin of the Atomic Scientists*, the organization of scientists that assesses the risk of catastrophe, today assesses that the nuclear risk has never been higher.

Let us have no illusions. An atomic war will kill hundreds of millions of people. Some, more fortunate, will be vaporized instantly, while the majority will die a little later, amid atrocious suffering.

The nuclear winter that will follow nuclear war – that is, the sudden climate change that follows nuclear explosions – will generate massive famines for those who have survived, and destroy civilization as we know it.

Does it seem reasonable to you that humankind, made up of people that are sometimes reasonable, like you, dear reader, and me, accepts such a level of risk?

1955: Voices of Reason

On 9 July 1955, Bertrand Russell and Albert Einstein, together with several colleagues, released a manifesto inviting intellectuals from all over the world to join a discussion on the risks to humanity posed by the existence of nuclear weapons.

> We are speaking on this occasion, not as members of this or that nation, continent, or creed, but as human beings, members of the species Man, whose continued existence is in doubt. The world is full of conflicts [. . .] All, equally, are in peril, and, if the peril is understood, there is hope that they may collectively avert it.
>
> We have to learn to think in a new way. We have to learn to ask ourselves, not what steps can be taken to give military victory to whatever group we prefer, for there no longer are such

steps; the question we have to ask ourselves is: what steps can be taken to prevent a military contest of which the issue must be disastrous to all parties?

Inspired by the Russell–Einstein manifesto, in 1957, in the village of Pugwash, Nova Scotia, Canada a group of scientists met and founded the Pugwash Conferences on Science and World Affairs, which, almost forty years later in 1995, was awarded the Nobel Peace Prize. Pugwash's goal is:

> to bring scientific insight and reason to bear on threats to human security arising from science and technology in general, and above all from the catastrophic threat posed to humanity by nuclear and other weapons of mass destruction.

Pugwash's historical influence has been remarkable. Its contribution to the push towards the nuclear arms reduction agreements of the 1990s has been acknowledged by top-level politicians, for example the former US Secretary of Defense Robert

McNamara on one side, and President Gorbachev on the other.

The most effective of these agreements have been the START I and II treaties (Strategic Arms Reduction Treaties). These are bilateral agreements between the United States and the Soviet Union that have led to a drastic reduction and a limitation of strategic offensive weapons. The START I Treaty was signed on 31 July 1991 and came into force on 5 December 1994.

Proposed by US President Ronald Reagan, the START treaty in its final implementation at the turn of 2001 led to a formidable result: the removal of about 80 per cent of all strategic nuclear weapons then in existence. The former Soviet Union's nuclear weapons stockpile, for example, has fallen from 12,000 to 3,500. That is still too many, but better than the madness of 12,000 atomic bombs. Sometimes human beings, even politicians, behave reasonably.

After nearly 100 million people being massacred during the two world wars, and after the horror of Hiroshima and Nagasaki, humankind tried to bring back reason. This led to organizations such as the International Atomic Energy Agency; the dream of the United Nations, where all countries together renounce war and agree to collaborate to prevent it; international research centres such as CERN in Geneva, open to all the nations of the world; and a wealth of supranational institutions, meant to play the pacifying role that governments have within states. All this was greeted with enthusiasm by the suffering humanity that emerged from the Second World War, mourning tens of millions of deaths.

Then the Cold War kept the world in suspense, hanging onto the nuclear risk, and was studded with local conflicts where the two superpowers clashed, each strongly invested in a powerful ideology, with opposing dreams about how to organize the future, but wisely managing to avoid a direct nuclear dogfight.

The internal collapse of the Soviet project, which did not withstand external pressure and structural weakness, left the United States in a dominant position. In the first years of its undisputed global empire, the United States sought the consensus of the planet, respecting legality and international institutions. When George H. W. Bush attacked Iraq – for better or worse – he did so with the approval of the United Nations, gathering a vast coalition of countries, convinced – rightly or wrongly – that Iraq's aggression against Kuwait had to be forcibly repelled by the entire planet, under the wing of international institutions. The world was embarking on a serious attempt to implement legal dispute management, as within states.

But it didn't last long. The intoxication of power ended up prevailing. When George W. Bush attacked Iraq again, he did so against international law, without the approval of the United Nations, even defying the pleas not to do so from allied countries such as France. Italy, my country, followed up in the demolition of international law by tagging along with the winning power of the moment, as it did forty-three years earlier with Germany. And so did others, like the UK, maybe nostalgic for its lost empire.

When NATO bombed Serbia, it did so against international law, without the support of the United

Nations, with a substantial part of the world opposing it. Propaganda managed to convince many that bombing Belgrade was altruism (as many in Russia are convinced that Russia fights in Ukraine out of altruism).

Instead of leading towards shared peace, upholding the international law that it had helped to establish in the post-war period, the United States has chosen to put its own empire first, to privilege violence, and to export continuous wars around the world. Thirst for domination has prevailed. The US, never attacked, has been at war almost continually since the end of the Second World War, constantly outside international law. A few countries, including mine, have always closely followed the master. This evolution became complete at the end of 2025, with the publication of the US National Security Strategy document, which unambiguously states that the US government opposes supranational institutions such as the United Nations and the International Court of Justice.

Russia, weaker, has also started, not as many but not a few, wars and invasions, from Afghanistan to

Georgia to Ukraine, likewise justifying all this in terms of generosity or defence.

Only China, among the great powers, has limited itself to very few clashes outside its borders recognized by international treaties. Western propaganda is unleashed on China's policies towards Hong Kong, Taiwan, Tibet, or the Uyghurs (which, under international law, are Chinese regions) without ever daring to compare the number of deaths in these regions with the millions killed by the US government and its allies *outside* their borders.

No one is a shrinking violet. Violence and thirst for power have always been widespread in the world. But in the world, and in all countries, there have always been great forces that have worked with realism and idealism together, to build a peaceful coexistence, to lessen the violent oppression of one over the other. For justice, equality, true democracy. I would like these forces to prevail.

To contribute to a less violent and less unjust world, the first thing we must do is abandon the fairy tale that our Western countries have been and are today on the side that contributes to reducing violence, suffering and risks. Because of a combination of short-sightedness, petty convenience, weakness and laziness, political choices and the consequent orientation of media propaganda in Western countries have been, and are today, part of the problem, not part of the solution.

1964: The Chinese Bomb

During the first Taiwan Strait Crisis, on 12 September 1954 the US Joint Chiefs of Staff recommended the use of nuclear weapons against China. On 6 March 1955, Eisenhower openly threatened to use nuclear weapons against military targets of the People's Republic of China in Fujian. At the end of March, Admiral Robert Carney announced that Eisenhower was planning 'to destroy the military potential of Red China'. Not surprisingly, shortly afterwards Mao Zedong began China's nuclear programme. As always, we become aggressive for fear of the aggression of others.

China exploded its first nuclear bomb ('596') in 1964 at the Lop Nur test site in the remote northeast of the country. Just two years later, it had a fission bomb capable of being installed on a nuclear missile.

China joined the Nuclear Non-Proliferation Treaty as a nuclear power in 1992.

As of February 2024, China has an estimated total inventory of about 500 nuclear warheads. Its nuclear arsenal is expected to continue to grow over the next decade. Projections suggest that it will deploy more or less as many intercontinental ballistic missiles (ICBMs) as Russia or the United States at the end of this period, or sooner. Western forecasts of China's economic growth have been regularly under-estimated for decades. In fact, Western media has been talking about the 'coming end of China's growth' for decades, while China keeps on growing.

As noted earlier, China's transformation into a major power greatly complicates deterrence. As in the three-way 'duel' at the end of Sergio Leone's *The Good, the Bad and the Ugly*, the game of who shoots who first, who allies with whom, and who trusts whom and who is cheating whom, becomes like a poker game. The simplicity of two-way nuclear deterrence is lost. If I am A and B attacks me, and I destroy B, I only make C win. What should I do? To do everything so that B and C destroy each other? Create a trap for this to happen?

China is the only one of the nuclear signatories to the Non-Proliferation Treaty to have explicitly adhered to the 'no first use' policy, officially declaring that it will never use nuclear weapons against any other state unless it is attacked with nuclear weapons. (India has an equivalent but somewhat more ambiguous policy.)

Similarly, China officially follows the doctrine of the 'least deterrent', creating a nuclear arsenal officially aimed exclusively at acting as a credible deterrent against a nuclear enemy attack. NATO nuclear countries and Russia, on the other hand, reserve the right to use nuclear weapons even in response to a conventional attack (which in the case of Russia is currently happening).

2003: Libya Renounces the Bomb, and it Ends Badly. North Korea Does Not, and No One Touches It

Then there are the 'small' atomic nations. To understand the logic of these, which is often presented incorrectly, I find a parallel between the development of the North Korean and Libyan nuclear programmes to be illuminating.

Both countries come into conflict with the United States and its allies. North Korea was born from the end of the Second World War, and from the military competition between the Soviet Union and the United States to divide the planet between them. Korea remained broken in two, like Germany.

The Korean War originates (once again) from

a misunderstanding, or rather from two misunderstandings. First, Soviet intelligence was convinced that the Americans would not defend South Korea against an attack from the North. Based on this information, Moscow gives the green light to Kim Il Sung, who rules North Korea, to invade South Korea and reunify the country. This is the first misunderstanding: the assessment of the Soviet intelligence is wrong. The Americans intervene and repel the attack.

Next, at this point it is the US intelligence that gets it wrong. It is convinced that China would not intervene, and the US military has the green light to invade North Korea. This is the second misunderstanding: the American military finds an army of 2 million Chinese soldiers facing them as soon as they invade the North (soldiers whom the Chinese government formally considers 'volunteers').

In the end, we are back to square one, minus 3 million dead young men, on both sides. No results, either on one side or on the other, for two wrong calculations. (Precisely as now in Ukraine. The Russian leadership probably thought it was easy to control Ukraine and did not anticipate the massive Western military support: hence the mistake. On the other hand, Ukraine's Western allies boldly expected that their weapons and 'devastating economic sanctions' would cause Russian aggression to fail – another mistake. Mistakes paid for with death after death and destruction after destruction. The most immoral of all are those pushing for the continuation of the war from far away.)

As a result of the events of the Korean War, the political leadership in North Korea, formally still at war with South Korea, has been under constant Western threat. For instance, it has had to deal with regular joint American and South Korean military exercises off its coasts.

The leadership soon realize that nuclear weapons are life insurance against a further American attack, when its historical ally, Russia, is weakened. It manages to obtain nuclear technology, saving the country from invasion and destruction.

In Libya, Muammar Gaddafi is less far-sighted. He too finds himself in conflict with the West. The reason is different. Gaddafi works hard to build pan-African alliances to free Africa from suffocating post-colonial subjection, which continues (to this day) to suck up the continent's resources. Gaddafi is strong because of Libya's abundant oil resources, and judges Africa to be weak because it is divided. He strives, thanks to the resources of his country, to co-ordinate African states and create a common defence against Western interference. This is intolerable for the West, which does everything to destroy him, and whose propaganda describes him (as always with its enemies) as a ridiculous madman. The West succeeds in isolating Libya and reducing it to a pariah country.

Gaddafi engages in a nuclear programme to defend himself. But the entire powerful West is against it, and in a difficult moment, under accusation of terrorism (the 'terrorists', by definition, are any adversary who dares to use force: Garibaldi, the Italian independence hero, was the 'terrorist' par excellence, seen from Austria. Force, according to everybody's propaganda, is only exclusively legitimate for 'our' side), Gaddafi ends up negotiating and agrees to abandon the nuclear programme in exchange for assurances that Libya is readmitted to the assembly of nations.

Never trust the West. Without nuclear weapons, Libya has no deterrence. Gaddafi's house is bombed, his children killed, his country, prosperous thanks to oil, is devastated. He is finally killed. His country, decades later, is still in chaos, with immense suffering for everyone.

For the Libyans, it would obviously have been better to have a minimum of nuclear deterrence to defend themselves, as North Korea has managed to do, from the overwhelming imperial power that does not tolerate those who do not submit.

Why are these simple considerations relevant? Because we often hear that the risk of nuclear war depends on the fact that there are small countries that have atomic weapons. This is false. The problem of nuclear weapons is not the small countries, such as North Korea or Libya. For the governments of weak countries, an atomic bomb is simply life insurance, never to be used, except under direct attack by others.

In the West, propaganda trumpets the risk that North Korea could drop a nuclear bomb on Los Angeles, omitting to say that this could only happen as a reaction to the US attacking North Korea. It is obvious that no ruler of a small country would be the first to use a nuclear weapon: if they did, the leader would end up vaporized 15 minutes later. The only time they could think of doing so is for revenge, when all was lost. The reason the US doesn't tolerate North Korea's nuclear bomb is not because of the risk of being attacked: it's because it prevents the US from attacking North Korea. It limits its power.

The problem of nuclear weapons is not small countries. The problem is the large countries, for which atomic bombs are instruments of domination, and the temptation to use them beyond deterrence – and the risk of an error and an escalation into a nuclear conflict – is incessant and growing.

The Twenty-first Century in Brief

2005: In an interview, the former President of the Italian Republic Francesco Cossiga reveals that, during the Cold War, Italy's US atomic bombs had the 'tactical' objective of destroying Prague and Budapest.

2019: The United States withdraws from the INF Treaty, which bans intermediate-range nuclear weapons. This step dramatically increases the risk of first use. All war simulations show that, while strategic intercontinental weapons can contribute to deterrence, intermediate-range nuclear weapons reduce the time available for deterrence to work, defeating its logic.

2023: On 21 February, Russia suspends its participation in New START. It does not withdraw from the treaty and states that – for the time being – it continues to respect the numerical limits of the Treaty. On 2 June, the United States revokes the visas of Russian nuclear inspectors. The treaty no longer exists. Of course, everyone blames each other. One of the fundamental building blocks of the logic of deterrence is in fact no longer there.

2025: In a rare positive moment for the planet, President Trump proposes negotiating with China and Russia to limit nuclear weapons and reduce the military spending of the three countries by 50 per cent. Perhaps the Good Lord enlightened him for a single moment. If so, Good Lord, please do it again, enlighten him and the other political leaders of the planet. We humans alone do not seem able to avoid always falling back into wars.

2025: This is the text of the *Bulletin of the Atomic Scientists*, who publish what is called the 'doomsday clock', an ongoing risk assessment:

In 2024, humanity edged ever closer to catastrophe. Trends that have deeply concerned the Science and Security Board continued, and despite unmistakable signs of danger, national leaders and their societies have failed to do what is needed to change course. Consequently, we now move the Doomsday Clock from 90 seconds to 89 seconds to midnight – the closest it has ever been to catastrophe. Our fervent hope is that leaders will recognize the world's existential predicament and take bold action to reduce the threats posed by nuclear weapons, climate change, and the potential misuse of biological science and a variety of emerging technologies.

In setting the Clock one second closer to midnight, we send a stark signal: Because the world is already perilously close to the precipice, a move of even a single second should be taken as an indication of extreme danger and an unmistakable warning that every second of delay in reversing course increases the probability of global disaster.

2026 Update: On 27 January 2026, the Doomsday Clock is set at 85 seconds to midnight, the closest the Clock has ever been to midnight in its history.

The Bad Conscience of Physicists

Many physicists, colleagues of mine, have tried to come to terms with the poisoned chalice that their discipline has offered to humanity.

As in all professions, scientists are different from each other, each with their own sensitivity, their own reading of the world and their own political ideas, and each reacts in their own way to the awareness of being part of the guild that has given humanity the prospect of burning ourselves alive and the risk of the end of civilization (since this is what we are talking about). Many physicists have felt and feel the weight of this responsibility. This book is also an expression of this weight.

Many physicists I admire have reacted by writing and speaking, engaging in organizations such as

Pugwash, or acting on their own, to alert humanity to the tremendous risk of a nuclear war, to sound the alarm, to try to divert humanity from sleepwalking towards the abyss dug by our discipline.

One name among many: Otto Hahn, who had discovered the barium in uranium irradiated with slowed neutrons, opening the way to the understanding of nuclear fission. After the war he became director of the Max-Planck Institute (the post-war heir of the Kaiser-Wilhelm-Gesellschaft) and took a decisive stand against the use of atomic energy in warfare. He denounced the use of science for the construction of weapons in times of war as a perversion of science, and he pushed the planning of the country's scientific efforts in the direction of social responsibility.

These efforts have not been in vain. For example, as I mentioned, the commitment of Pugwash and many Western and Soviet scientists played an important role in the agreements between the United States and the Soviet Union that drastically reduced the risk of a nuclear confrontation (agreements from which the United States and Russia have now exited).

Other physicists, tormented, have sought justifications. The justification that horrifies me the most, and which I sadly heard in person from the scientist of the previous generation that I most admire intellectually, John Wheeler, is that dropping the bombs of Hiroshima and Nagasaki 'saved the lives of American soldiers'. Just mentioning it here, I am overcome again by the sense of nausea I felt when John said these words as we were walking in the Princeton countryside. The moral justification for burning alive hundreds of thousands of non-Westerners is that it enables the invasion of a country on the opposite side of the planet without risking the lives of young Americans.

This was the logic of centuries of European colonialism, and it was also, and still is, the logic of so many Western wars during the so-called post-war period. A living Westerner is worth how many dead non-Westerners?

It is a logic that we still see at work every day. How many Palestinians killed weigh as much as a single Israeli killed? I find this logic repulsive, impregnated as it is with racism, violence and contempt for humanity.

I often wonder, too, if most of the people on the planet, billions of human beings who almost all have, in one way or another, forebears who in this logic have been massacred, or reduced to slavery, will ever forgive the colonial and post-colonial powers that we are. The problem of the present is not the past, for which we who live today are not responsible: it is the future, for which we are.

The most common justification for building the super-bomb has been that it contributed to the defeat of Nazism. It is a justification that does not hold: Nazism capitulated before the super-bomb. Then, since Germany was not building the bomb, the justification becomes that Germany 'could' have built it.

Many physicists ostracized Werner Heisenberg in the post-war period. I want to say a few more words on this, because this debate has muddied the waters.

Heisenberg, in his fundamental simplicity, is nevertheless a figure who confronts us with general questions. He never joined the Nazi Party or affiliated organizations. He was among the few university professors not to sign a manifesto of support for Hitler. After receiving the Nobel Prize, he refused to attend a rally in honour of Hitler in Leipzig, his hometown. At the time, such a choice could have cost him his life. His friend and collaborator Weizsäcker, of very different social background, came from the old Prussian military aristocracy, which never digested Hitler. Heisenberg fiercely and openly opposed the ideology of the 'Aryan Science' of Nazi physicists such as Philipp Lenard and Johannes Stark, defending Jewish colleagues to the point of being accused by the Nazis of being a 'white Jew', i.e. a German who sold out to the Jews. This resulted in him being investigated by the Nazis.

Yet, despite all this, Heisenberg is repeatedly accused of Nazism, even by colleagues whom I respect. And during the writing of this book, I met colleagues who told me about Heisenberg's Nazism. Why? The answer is important.

First, because he did not want to leave Nazi Germany. He continued to work for his country and his government during the war, and to represent Nazi Germany when he travelled to occupied countries. This is enough to discredit a person in a world where you are either on one side, or you are on the other. Where those who do not fight with you are seen as your hated enemy.

Heisenberg never saw things in these terms. This made him intolerable and still arouses the contempt of those who divide the world into absolute criminals and absolute champions of justice. Heisenberg, as he has been described by those who knew him, saw the world differently, as a mixture of good and not-good things. The post-war period did not accept such complexity. Victory was to be read as the triumph of absolute good over absolute evil. How else could humanity tell itself that it had just killed 70 million people?

Heisenberg was certainly not a coward. On the contrary. He was courageous to the point of recklessness. Before the war he could have emigrated to the United States with his family, but he chose to stay in Germany, where he might have been sent to the front and could easily have died. At the end of the war, he could have simply waited for his arrest by the Allied forces without facing any danger; instead, he ventured on a daredevil bicycle trip through burning Germany to see his family again. He was even a member of the 'Wednesday Society', a club for the German elite, many of whose members took part in the failed assassination attempt on Hitler on 20 July 1944 and were executed.

He hated Nazism, but he loved his country. A combination that today is intolerable to many.

But maybe there is something much deeper in the antipathy of many physicists toward Heisenberg: the fact that German scientists did not engage in the super-bomb remains intolerable for many physicists in the victorious countries, because it invalidates their moral justification for creating the horror. The collective sense of guilt perceived by many physicists has to confront the disconcerting fact that the physicists of the evil regime par excellence remained immune to it.

Heisenberg was accused of everything. Of flaunting moral superiority for not having made the bomb. Of having lied by saying that he could have made the bomb but had not wanted to. Of having left Germany without a bomb because he had miscalculated the amount of uranium needed for it. As if a single calculation by a single person could be relevant in such a situation . . .

Rather than acknowledging the fact that the super-bomb was not needed to win against Germany, the debate shifted to the hidden reasons for Heisenberg's actions, as if these were relevant.

After the war, Heisenberg was looked upon with suspicion by those who had once been his friends. Attempts to talk to and understand each other did not work. The wounds caused by the immense tragedy of the war were too deep. Perhaps the vague sense of guilt felt by many physicists could not tolerate the obvious innocence, in the construction of the worst weapon of mass destruction ever produced by humanity, of the greatest German scientist, who had the defect of loving his country and had worked for Hitler.

After the war, along with Weizsäcker, Heisenberg was the driving force behind the 1957 Göttingen Manifesto, which called on the West German government to abandon its plans to develop its own nuclear weapons, anticipating the subsequent anti-nuclear peace movement. He was committed to international scientific co-operation, playing an important role in the creation of CERN in Geneva in the 1950s. During his tenure as president of the Alexander von Humboldt Foundation, from 1953 to

1975, he defended the ideal that scientists should form an international family characterized by lasting ties.

What had Heisenberg said to Bohr, or what would he have wanted to say, in the fateful meeting in Copenhagen, at the beginning of the war, when Germany, before attacking Russia, seemed invincible?

Perhaps, as Bohr later argued, that Germany was winning and it was better to come to terms with this fact. Or, as Heisenberg himself later suggested, would he have wanted to discuss the moral implications of building a super-weapon in wartime? Or, as Elisabeth, his wife, would write, did he signal to Bohr that Germany would *not* build the bomb? Or, as Weizsäcker – a more political animal, and probably the one who had pushed for this meeting – wanted, did he to suggest that an open channel of communication between the physicists of the two sides should be maintained?

Does it matter what they actually said to each other? The emotions, the short time available, must have been too overwhelming for both in the heart of a tragedy. And certainly, they were both incapable of living the new reversed roles in which they had found themselves. Heisenberg was politically naïve; Bohr was overwhelmed by anti-German emotions. Neither of them was political at heart. The tragedy is not in what they said to each other.

The tragedy is in what they could have said to each other if they had been up to it and what they did not do: listen to each other, try to understand each other, and believe in each other's sincerity, which could not be doubted. Perhaps, if the two greatest scientists of quantum physics had managed to do so, the horrendous misconception of the United States believing that Hitler was on the verge of having the super-bomb would not have taken hold. Bohr would not have arrived in the US with a sketch by Heisenberg convinced, mistakenly, that it was a plan for the bomb. Nuclear energy, developed in peacetime, could have had the possibility of being kept under common control, instead of threatening us as it does. These are dreams, obviously. History is not made up of 'ifs'.

But the future, on the other hand, the future *is* made up of 'ifs'. Different possible futures are open. They depend on our choices today. Choices that, I believe, we are making wrongly.

★

There are also physicists who have preferred to ignore the issue entirely: both the moral question and the political question, even among those who have been deeply immersed in it.

I don't allow myself to judge anyone. As far as I am concerned, everyone deals with their own conscience. But among the different positions and commitments that my colleagues have taken today, there are some that I admire and that I think are useful to humanity, some whose torment I understand, and others I don't admire. Among these latter there are physicists, even those whose work is of great scientific value, who have heightened the risk for humanity and yet have not left any sign of concern in this regard.

Some have done so perhaps out of simple love for their country, committing what I believe is an error of judgment and a moral error: putting the good of one's own country above the common interest of humanity. Risking collective catastrophe in order to increase the power of one's own country. Patriotism has caused more corpses than the Black Death. Many are making this moral mistake today.

Others have done so simply to avoid asking

themselves any questions, interested only in their own science, without dealing with anything else. Among these I think is Enrico Fermi, with whom I opened this book. Fermi was part of a very small group of scientific giants, which includes Einstein, Paul Dirac, Bohr and few others. As far as I know, he did not publicly express any political or moral opinions related to his work in science.

In 1934, he was a member of the Fascist Party, and had been appointed by Mussolini as a member of the Royal Academy of Italy. He was certainly not a fanatical Fascist, but he adapted to the situation, participating in official ceremonies in uniform. The racial laws put him in difficulty because his wife was Jewish. He did not react by protesting. He took advantage of the trip to Stockholm to collect his Nobel Prize and escaped with his wife to the United States. As soon as he arrived in Sweden, knowing that he would not return to Italy, he stopped wearing the uniform that the regime wanted him to wear and did not perform the fascist salute that the regime asked of him, the salute that he had given until the day before. From Stockholm he went directly to live in the US, where he became a good citizen, as he had been a good citizen of Fascist Italy, and participated in an important way, from Chicago, in the Manhattan Project to build a bomb that might, had events unfolded differently, have been used against his native country.

On 2 May 1945, President Truman formed an interim committee with an advisory function with respect to the use of the atomic bomb, which was now ready. The committee was composed of eight people, supported by an advisory scientific committee comprising Robert Oppenheimer, Ernest Lawrence, Arthur Compton and Enrico Fermi. In a report dated 1 June, the committee concluded that the bomb should be used as soon as possible, against a war facility, *surrounded by workers' homes*, and that *no warning or demonstration should be given*. I don't know what Enrico Fermi argued within the committee. But in the years since, as far as I can find, he did not publicly express any sense of unease at having participated in recommending the extermination of innocents.

For me he remains a great scholar, whose science I admire deeply – so much so that I decided to name a programme of important scholarships that I manage after him. But if I had children, I wouldn't be happy if they behaved like that in life. When I said this in Italy it caused a stir, because Fermi is a national hero. I am not writing this for the sake of controversy. I write this because I think that today we desperately need everyone, and all of us scientists in the first place, not to remain locked in our own worlds but to work to stop the current race to war. To avoid millions of humans being burned by nuclear fire.

Final Words

The day before an earthquake, no one thinks about earthquakes. My father told me that the week before the final collapse of the Republic of Salò, the last incarnation of Italian fascism, the parliament had voted to reform the postal service. As if everything were to continue the same for ever.

The day before the assassination of Archduke Franz Ferdinand and his wife, Sophie, no one imagined the First World War. The great powers were arming themselves more and more (as today) based on the idea that 'if you want peace, prepare for war' (as today); they declared themselves honestly committed to defending their allies (as today), and embraced the idea that if you are well armed and have good alliances, you don't make war (as today).

Humans have often sleepwalked into cata-strophic wars, in periods (like today) of mutual fear, increasing belligerence, and heavy investments in armaments, mistakenly thinking that they are avoiding the worst. Our politicians are not far-sighted; they think locally and in the short term, just as the politicians of the past. The impressive array of miscalculations I have listed on the previous pages shows how irrational the art of politics is, and how unreliable the decision-making elites are. But now we have a nuclear catastrophe looming.

The war propaganda of today, as I am writing these lines, focuses on the misleading example of the days before the Second World War. We often hear the argument that the Second World War broke out because of the *appeasement* of Chamberlain, the British prime minister who signed the Munich Agreement of 1938, accepting the annexation to Germany of the German-speaking Sudetenland region of Czechoslovakia. According to this logic, this was the reason for the Second World War. It would follow that appeasement, or finding agreements, instead of resolving issues by force, leads to war.

The argument does not stand up, and it is dramatically dangerous. The Second World War broke out for a variety of situations and reasons, including the glorification of war by Fascist and Nazi ideology, the rearmament of Germany, the class struggle within many countries, the fear of Soviet communism, the desire for revenge of a Germany humiliated by the First World War, strengthened by an explosive industrial revolution, just out of a devastating crisis, and many other reasons. Among these reasons there is also the *lack* of an agreement between Poland and Germany to resolve the 'German issue' there. Let us not forget that at the time the 'German issue' was declared legitimate by all the powers. England *failed* to convince Poland to accept an agreement. If one emphasizes the diplomatic failure of the negotiations between Germany and Poland, rather than the success of Chamberlain's appeasement, one could argue that the war broke out because there was *no* appeasement, despite many considering it reasonable. But things are always complex, of course. War is not avoided by searching for simplistic historical analogies.

Today's Russia is different from Nazi Germany for countless reasons. Europe has no real reason not to live peacefully from the Atlantic to the Urals. On the contrary, it has obvious reasons for doing so, as years of past German foreign policy, opposed by the UK, showed. Similarly, the whole world has many more reasons to collaborate rather than to wage war against each other. Many countries around the world stand by this.

Russia has no intention, nor way, to invade Paris, Rome, Berlin or London. There are problems, for example in the Baltic. But reasonable beings find reasonable solutions – if they don't try to resolve everything with dreams of military supremacy, and if they behave like rational beings instead of gorillas. Russia has no military capacity to invade Western Europe, but it has total deterrence against attacks: 4,000 nuclear warheads. It is not and has no intention or way of being an enemy of Europe, except to the extent that Europe, and the United States and the United Kingdom in particular, turn it into one.

What we need today is to learn not to close our eyes and not to go with the flow that is leading us towards the abyss. We need everyone's commitment, including intellectuals and, among them my colleagues, the scientists of the discipline that created the worst weapon of mass destruction.

Individual politicians, newspaper editors, journalists and television personalities focus more on the effectiveness of their positioning in the short term than on the collective good in the long term. A great Italian politician responded to a proposal for a balanced global reduction in military spending, signed by sixty Nobel laureates, with the words: 'Yes, it is right and far-sighted, but it does not bring votes.'

This is the problem with democracy.

I would like countries to find the wisdom to bring a little rationality into the current madness of international politics made up of wars, misunderstandings and power games that end badly. We have seen in the previous pages how many countless corpses have been left on the ground due to misunderstandings and short-sightedness. I would like some of our politicians to be able to look a little beyond their noses, to think about the fate of the planet and take a stand against the current arms race and suicidal belligerence.

The only reasonable response is to stop looking at enemies as enemies, as we are doing today. Stop thinking that they must submit to our cultural, economic and military supremacy. Stop thinking that the problem is how to be stronger and more aggressive than others.

The problem today, the days of nuclear risk, climate crisis, still widespread poverty, increasing belligerence, is not who wins. The problem is how to avoid war. To learn to collaborate instead of fighting. If we think we can avoid war by becoming stronger and more aggressive, looking at others in disdain, the result is that others try to become even stronger and more aggressive than us, and to look at us more and more in disdain.

I believe that we must learn to see enemies as human beings who frighten us because they are afraid of us. This is not cheap moralism. It is being rational instead of irrational. We are behaving like neighbours who come to blows because of a hedge placed one metre to the right or to the left. We do it armed with thousands of nuclear missiles pointed at our heads, just a touch of a button from being launched.

We will certainly arrive at a reasonable and effective shared political structure for the whole planet. The uncertainty is whether we get there before or after the nuclear catastrophe. It depends on us.

Are there any adults in the control rooms?

A Brief History

1900: Max Planck writes the work that paves the way for investigating quantum physics.

1904–5: Russo-Japanese War.

1913: First quantum theory and model of the atom by Niels Bohr.

1914–18: First World War.

1917: Russian Revolution.

1918: Independence of Poland, Russia, Austria-Hungary and the German Empire.

1919: The Weimar Republic is born in Germany.

1919: Mussolini founds the Italian Combat Leagues.

1922: March on Rome by the National Fascist Party.

1925: Werner Heisenberg writes the key work for modern quantum theory.

1929: Inflation and stock market crash in the United States.

1932–45: Soviet–Japanese border wars.

1933: Adolf Hitler becomes Chancellor of Germany.

1934: Enrico Fermi breaks the atomic nucleus.

1935–36: Italy at war with Ethiopia and annexation of Ethiopia.

1936: Spanish Civil War, leading to the dictatorship of Francisco Franco.

1937–45: Second Sino-Japanese War.

1939: Non-aggression pact between Nazi Germany and the Soviet Union.

1939: German invasion of Poland: start of Second World War.

1939–43: Italian occupation of Albania.

1940: German invasion of Denmark and Norway; then Belgium, France, the Netherlands and Luxembourg. Germany and its allies control non-Soviet continental Europe.

1940: Italian bombing of Tel Aviv.

1941: Germany invades the Soviet Union, the largest invasion in history.

1941: Bohr–Heisenberg meeting in Copenhagen.

1941: Japan attacks the US naval base at Pearl Harbor and the United States enters the war.

1943: Fall of fascism in Rome, foundation of the Republic of Salò.

1943: The Soviet Union begins to repel the German invasion.

1944: The Americans arrive in Rome.

1945: The Russians arrive in Berlin.

1945, May: Surrender of Germany.

1945, July: First nuclear test (Trinity test).

1945, August: US atomic bombs on Hiroshima and Nagasaki.

1945, September: Surrender of Japan.

1945, October: The United Nations is officially born.

1946–54: Indochina War.

1948: Universal Declaration of Human Rights.

1949: The People's Republic of China is declared by Mao Zedong.

1949: First Russian atomic bomb.

1950–53: Korean War.

1955: Russell–Einstein Manifesto for Nuclear Disarmament.

1955–75: Vietnam War.

1962: Cuban Missile Crisis: The world comes close to nuclear war.

1964: China's first atomic bomb.

1975: Fall of Saigon: Vietnam victory over the United States.

1982: Ronald Reagan proposes a treaty for the reduction of nuclear weapons, later signed by the Soviet Union and the United States (START Treaty).

1989: Collapse of Soviet communism.

1991: Dissolution of the Soviet Union.

1995: Nobel Peace Prize to the Pugwash organization for work on nuclear disarmament.

1996–99: Kosovo War, Italy participates in the bombing of Belgrade.

2006: North Korea's first nuclear test.

2011: NATO war on Libya, in which Italy also participates, and subsequent destruction of the country.

2014: The war in Ukraine begins.

2022: Russia invades Ukraine and NATO heavily defends it by arming it.

2023: Russia and the United States exit the START treaties.

2025: Israel–Iran conflict and US bombing of Iranian nuclear sites.

2025: The *Bulletin of the Atomic Scientists* estimates that nuclear risk is at an all-time high.

2026: The *Bulletin of the Atomic Scientists* sets the Doomsday Clock at 85 seconds to midnight, the closest we've ever been to nuclear catastrophe.